INSIDE SCIENCE

DISAPPEARING PRAIRIES

Jane Kelley

CONTENTS

Prairies Before People 3

Lost Giants 6

Prairies Today 12

Disrupted Food Chains 14

Threats to Survival 16

Glossary 24

PRAIRIES BEFORE PEOPLE

Fifty thousand years ago the world was a very different place. People lived mostly in Africa and in Eurasia (Europe and Asia). Many lived in open grasslands—on prairies that had formed over thousands of years.

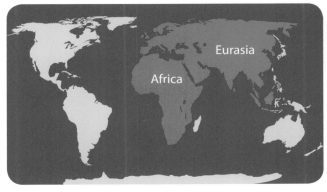

The green areas show where people lived 50,000 years ago.

If you could travel back in time to a North American prairie of 50,000 years ago, what would you see? You wouldn't see any people—but spread out before you would be enormous herds of mammoths, camels, mastodons, and other plant-eating animals (herbivores).

You would also see meat-eating animals (carnivores) like saber-toothed cats, short-faced bears, cheetahs, manned lions, and dire wolves hunting the herbivores. Many of these animals were so big that modern scientists call them megafauna.

mega (very large) + **fauna** (animals) = **megafauna**

Modern human | Saber-toothed cat | Stag-moose | Short-faced bear

North American megafauna about 50,000 years ago

Did you know that a beaver that was as big as a modern bear once lived in North America?

Mastodon

Mammoth

LOST GIANTS

As the world's climate became colder, people were able to move to previously **uninhabited** places. As more and more water was locked up in ice on the land, sea levels dropped. This opened up a land bridge that allowed people to move across from Eurasia into North America. As people moved into parts of the world where people had never lived before, including Australia and South America, they found prairies covered with megafauna.

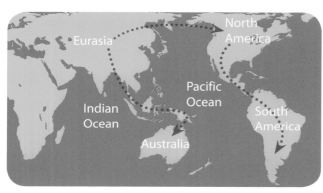

Migration out of Eurasia into North America, South America, and Australia

The tools these hunters used are called "the megafauna toolkit." They used this toolkit when killing and cutting up large animals.

Within 500 years of the arrival of people, most of the megafauna in North America disappeared. What happened to these animals?

There are several theories. The changes in the world's climate may have destroyed the **habitats** that these animals depended on. New diseases may have spread. There is also evidence that people hunted and killed them. After all, a huge animal would contain a lot of meat. These theories may all be true. As more evidence is gathered, other theories may emerge.

What happened in North America also happened elsewhere. People found megafauna living on the grasslands of South America and Australia, too.

Australian megafauna about 40,000 years ago

When people first moved into the Australian grasslands, they found giant kangaroos. At 7–10 feet (2.13–3.05 meters) tall, short-faced kangaroos were the largest kangaroos that have ever lived. Huge land crocodiles that grew up to 23 feet (7.01 meters) long hunted these giant kangaroos!

People did more than just hunt the animals they found living on the grasslands of Australia, North America, and South America. Sometimes, they started managing the prairie itself, changing how things worked. In Australia, for example, people have been burning off old grass for thousands of years. This kind of grassland management is called fire-stick farming. It means that what looks like a wilderness may not be quite as wild as it seems.

The result of fire-stick farming

! Did you know that some plants are adapted to fire? Fire and smoke help them to release their seeds.

PRAIRIES TODAY

Today, there are two main types of prairies, or grasslands. **Temperate** grasslands have four seasons—spring, summer, fall, and winter. **Tropical** grasslands have two—a dry season and a wet season.

Great Plains in North America

Prairies are called different things in different parts of the world.

Kirghiz Steppe in Central Asia

Mitchell Grasslands in Australia

> In the tropical grasslands of northern Australia, the wet season is called The Big Wet.

The world's prairies

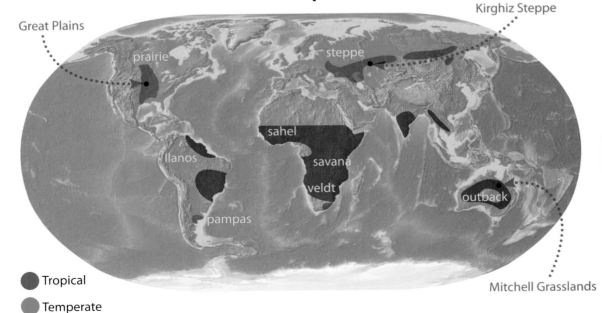

- Tropical
- Temperate

DISRUPTED FOOD CHAINS

Whenever herbivores eat plants and carnivores eat those herbivores, a food chain is formed. In the llanos, the tropical grasslands of northeast South America, there is a food chain that includes the world's largest rodent, the capybara, and one of the world's largest snakes, the anaconda.

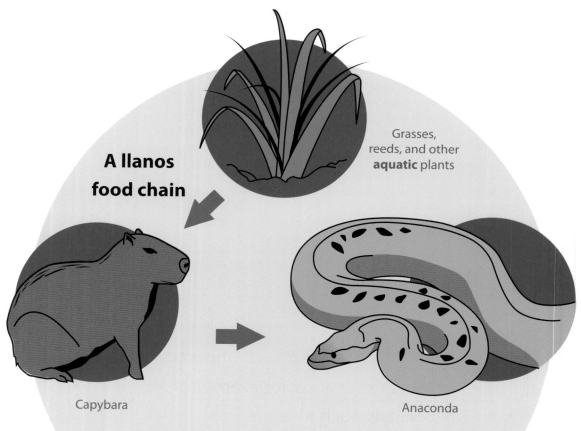

A llanos food chain

Grasses, reeds, and other **aquatic** plants

Capybara

Anaconda

During the wet season, anacondas move out into the water-soaked grasslands to hunt for capybaras. In the dry season, they hide in the streams and water holes where the capybaras drink.

Human
6 feet (1.82 metres) tall

Capybara
2 feet (0.06 metres) tall

Anaconda
30 feet (9.14 meters) long

The actions of people can have a huge impact on a grassland food chain. In parts of the llanos, for example, ranches are displacing wild animals, such as capybaras.

THREATS TO SURVIVAL

When ranchers and farmers started moving into the Great Plains in the 1800s, they introduced new animals and farming methods.

Ranchers brought horses, donkeys, cattle, and sheep to the prairies. Farmers brought plows that could cut through the dense roots of prairie grasses. It takes thousands of years for a prairie's topsoil to form. Plows turned this topsoil over in an instant, exposing the soil to the wind.

In the Great Plains, after several years of **drought** in the 1930s, the strong prairie winds simply blew the dry topsoil away. The worst affected area came to be known as the Dust Bowl. On April 14, 1935, it was so dark with blowing dust that people remember it as "Black Sunday."

Dust from the Dust Bowl was blown as far as Washington, D.C.

After Black Sunday, the ranchers and farmers adopted soil **conservation** measures, such as planting crops with deeper roots and **contour plowing**.

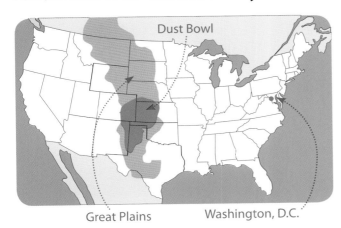

The Dust Bowl

Enormous fields planted with just a few crops cannot support the same food chains as the original prairie did. Without a **diversity** of plants, there can't be a diversity of animals. As a result, a few **destructive species** can take over.

Farmers control destructive crop-eating insects like the wheat stem sawfly with pesticides—poisons that kill insects. But pesticides don't just kill pests, they kill other insects, too.

> Did you know that when birds eat insects that have certain kinds of pesticides in them, they lay thin-shelled eggs that break before the baby birds can hatch?

The wheat stem sawfly can destroy a field of wheat.

Pesticide being sprayed

Because temperate grasslands don't have a wet season like tropical grasslands, they lack a **reliable** rainfall. Farmers store water in dams and distribute it through channels to **irrigate** their fields.

> The level of water in the Ogallala Aquifer has dropped by 50 feet (15.24 meters) over the past 40 years.

Another method of irrigation is to pump water up from an aquifer deep under the ground in a layer of **permeable** rock.

An aquifer

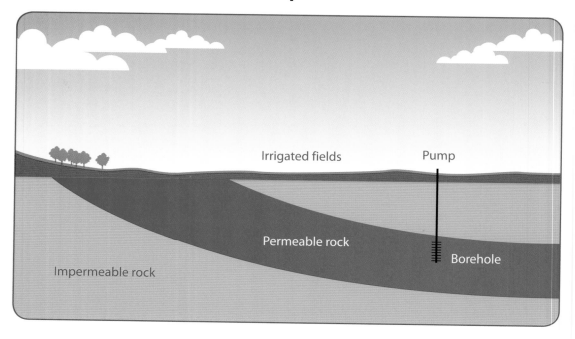

The Ogallala Aquifer, in North America, lies beneath 225,000 square miles (approximately 582,750 square kilometers) of prairie. It irrigates a huge area of land. The water in the Ogallala Aquifer trickled down through the permeable rock for millions of years. It won't last forever, now that it is being pumped out. Without water, the grasslands will die.

People are trying to save what's left of the world's grasslands, and the plants and animals that live there, before they disappear.

> Scientists estimate that prairie animal and plant species are dying out at an average rate of about one every 20 minutes.

A bison herd, Yellowstone National Park

The bison is an example of one success story. In the early 1900s, there were only a few hundred wild bison left. Thanks to conservation measures, there are now over 150,000 wild bison roaming the Great Plains.

Still, more and more of the world's grasslands disappear each year. Scientists calculate that by the middle of this century, over 50 percent of the grasslands in Africa, 90 percent of the grasslands in Australia, and much of the grasslands in North America, South America, and Central Asia will no longer exist.

It is estimated that an area of grasslands that adds up to the size of Texas turns into desert every year.

In too many parts of the world, prairies are shrinking and deserts are taking their place. If people make an effort, we can still save our prairies for future generations to enjoy.

GLOSSARY

aquatic—living in water

conservation—protection

contour plowing—plowing that follows the contour of the land instead of running in straight lines

destructive—harmful

diversity—wide selection or variety

drought—when rain doesn't fall for a long time

habitats—places where animals and plants live, such as prairies

irrigate—provide with water

permeable—allowing fluid or other substances to pass through

reliable—regular, unfailing

species—a single type of organism, for example, a coyote

temperate—having a moderate climate

tropical—having a warm, wet climate

uninhabited—unlived in, without people